GCSE Practice Papers
in Mathematics
(Intermediate Level)

J.J. McCARTHY BEd (Hons)

DP PUBLICATIONS LIMITED
Aldine Place, 142/144 Uxbridge Road
Shepherds Bush Green
London W12 8AA

1989

ALSO AVAILABLE:

GCSE Mathematics (Higher Level)

A CIP catalogue record for this book is available from the British Library.

ISBN 1 870941 30 6

Copyright J.J. McCARTHY ©1989

Typeset by
Alphaset
65A The Avenue, Southampton

Printed by
The Guernsey Press Co. Ltd.
Braye Road, Vale,
Guernsey, Channel Islands

Table of Contents

PREFACE

1. Aim

The aim of this booklet is to provide *sufficient* GCSE Practice Papers to enable teachers, parents and pupils to know how well prepared the pupils are for their GCSE Mathematics examination at Intermediate Level.

2. Need

The need was seen for a means of *testing* (both in the home and in the classroom) the *combination* of *breadth of knowledge* and the *speed needed* to convey that knowledge to the examiner.

3. Approach

There are *two* sets of *one hour* Practice Papers for *each* topic or group of topics (Area and Volume: Transformations etc.). Each set is followed by the relevant answers; teachers using the papers for class tests or for homework may wish to cut out the answers.

In all, there are *ten* pairs of topic-based papers *plus* two *mixed* topics for mock examinations/final revision.

NOTES:-

(a) The GCSE Boards' Syllabi, and their Papers, have all been closely studied and their main requirements incorporated.

(b) One hour was chosen as feasible in a *double period* at school and as a *reasonable time* to set aside for home self testing.

PART I

GCSE PRACTICE PAPERS

TIME ALLOWED
1 HOUR EACH PAPER

INTRODUCTION AND CONTENTS

Set aside an hour, *periodically*, to attempt each of the topic-based papers as you cover them in your studies. Check your answers *after* you have completed a paper.

In addition to the ten topic-based papers, there are two mixed example papers for final revision purposes. Each is designed to take about 1½ hours including time to read over your answers and correct your errors.

On the inside back cover is a formula sheet similar to one you will be provided with in your examination.

ALGEBRA AND GRAPHS
TIME 1 HOUR

1. Solve the equations:

   (a) $4x - 1 = 17$

   (b) $\dfrac{4x + 1}{3} = 17$

2. Draw a graph of the straight line $y = 6 - 3x$; let the line run until it intersects both axes.

   (a) Write down the co-ordinates of the points where it cuts

   (i) the x-axis

   (ii) the y-axis.

   (b) State the gradient of the line.

3. The cost, £C, of placing an advertisement in a local paper consists of a basic charge of £5 plus 50p a word.

   (a) Write a formula for the cost of printing n words:
   C = ...

   (b) How much would an advertisement of 15 words cost?

   (c) How many words were in an advertisement which cost £8?

   (d) Rewrite your formula in the form: n = ...

4. (a) Factorise $12xy + 24yz$

   (b) Remove the brackets: $4a(3b - 5c)$

   (c) Simplify: $5(x - 2y) - 4(2x - 3y)$

3

5. (a) Given that $c = \sqrt{a^2 + b^2}$,

    calculate c when a = 9 and b = 12.

    (b) Given that $y = \dfrac{x^2}{z^3}$, calculate y when x = −4 and z = −2.

6. For the relationship $\dot{y} = 3x^2$,

    (a) Copy and complete the table below:

x	−2	−1½	−1	0	½	1	1½	2
y				0				12

    (b) Using axes and scales of your own draw a smooth curve for the values (x,y).

    (c) Join the points (0,0) and (2,12) with a straight line; write down the gradient of this line.

    (d) Use your graph to solve the equation $3x^2 = 9$ (make clear your method).

7. The table below is for a set of points in a straight line:

x	1	2	3	4
y	5	a	9	b

    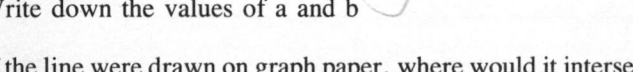

    (a) Write down the values of a and b

    (b) If the line were drawn on graph paper, where would it intersect the y-axis?

    (c) What is the gradient of the line?

8. Where do the lines x = 3 and y = x intersect each other?

4

AREA AND VOLUME
TIME 1 HOUR

1. A garden pond is circular with diameter 15m; it is surrounded by a concrete walk ½ m wide.

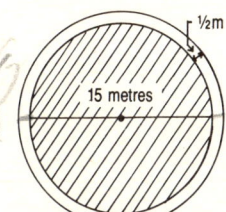

(a) State the radius of the outer circle.

(b) Find, by calculating the areas of the two circles, the area of the concrete to the nearest m².

2. A can of Quench has a base radius of 3½cm and is 12cm long.

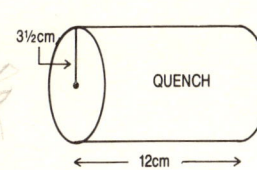

(a) Calculate the volume of the can to the nearest cm³.

(b) A larger can holds twice as much drink and is also 12cm long; calculate the base radius of the larger can.

3. To tile a part of their bathroom wall, Mr. and Mrs. Ali used square Mica tiles of area 49cm². The portion of the wall to be tiled measured 2.1m × 1.4m.

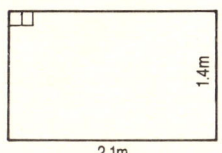

(a) How many tiles did they require?

MICA-TILES
£7.80 A DOZEN
10% OFF IF YOU
BUY OVER 500 TILES

(b) They saw this advertisement for the tiles in their local paper; How much did it cost to tile the wall?

4. A box of Sparker matches has dimensions 5cm × 3cm × 1.7cm as shown in the diagram; one of the open ends is shown by the clear side in the diagram.

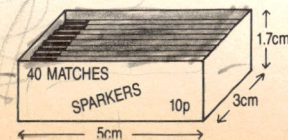

(a) Calculate the amount of cardboard required to cover a box, assuming both ends are open.

5

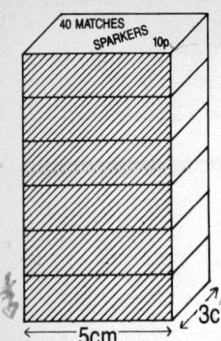

40 MATCHES
SPARKERS 10p

5cm
3cm

(b) The boxes are packed in sixes before being sold to shops; assuming they are packed face-to-face, draw a sketch of the packet and mark in its dimensions. Calculate the minimum amount of wrapping paper required to cover the six boxes – ignoring overlaps. You are advised to set your answer out as follows:

Front requires: __51__ × _____ = _____ cm^2
Top requires: etc.

5. The plan shows a racing track with the diameter of the semi-circular ends x metres.

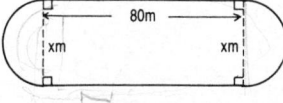

80m
xm   xm

(a) Taking one complete lap of the track to be 317m, calculate x to the nearest m.

(b) Using your answer to (a), calculate how long it takes a workman to dig up the area enclosed by the track, if he digs 48m^2 an hour. Give your answer to the nearest hour.

6. A student of architecture was told to make a 1:80 model of the end of a building; the triangle represents the model.

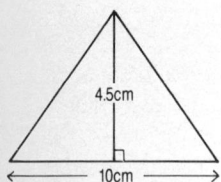

4.5cm
10cm

(a) What was the area of the end of the building in m^2?

(b) In making her model she was allowed to be out ±1mm in her drawing; what was the minimum possible area of her model?

7. In the diagram the area of the parallelogram ABCD is 32cm^2.

A   D
4cm
B   C E
11cm

What is the area of the triangle DCE?

8.

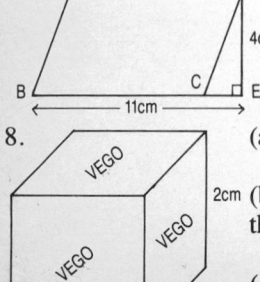

VEGO
VEGO   VEGO
VEGO
2cm
2cm
2cm

(a) What is the volume of one Vego cube?

(b) What is the smallest number of Vego cubes that can be put together to make another cube?

(c) In a wrapped cubical packet there are 343 cubes; what is the length of the edge of the packet?

CONSTRUCTIONS
(including bearings and scales)
TIME 1 HOUR

1. (a) Draw an equilateral triangle of side 5cm.

   (b) What is the height of the triangle?

2. The diagram is taken from a Thomson Holiday brochure for the South of Spain;

(a) Using the scale on the map, estimate the distance from Cordoba to Granada.

(b) What is the bearing of Seville from Jerez?

(c) What is the bearing of Malaga from Granada?

3. The positions of 3 trees T_1, T_2, T_3 in a local park are given in the diagram; T_2 is due north of T_1.

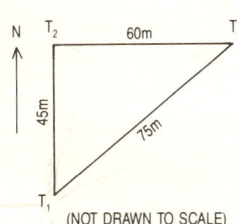

(a) Using a scale of 1cm = 7½m, draw an accurate diagram to show the relative position of the trees.

(b) What is the bearing of T_2 from T_3?

(c) The park authorities decided to plant a flower-bed within the triangle of trees; they instructed the gardener to make the edge of the flower-bed 13½m from the edge of the 'triangle'. Draw into your diagram the outline of the flower-bed.

4.

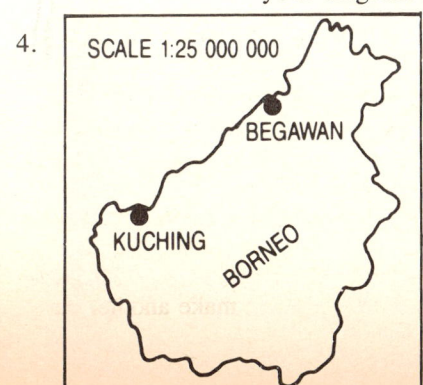

The map shows the outline of the island of Borneo with the scale inset.

(a) Calculate how far apart, in km, are Kuching and Begawan.

(b) Two other places, not marked on the map, are known to be 1150km apart; how far apart would they be on the map if they were shown?

5. A ship leaves port P and sails 600 km to port Q on a bearing of 120°.

(a) Using a scale of your own (which must be stated) draw an accurate diagram of the route.

(b) The captain's instructions are to patrol the area within 240km from Q, but not to cross Southwards over the imaginary East-West line through Q. Show clearly on your diagram the area to be patrolled.

(c) If the ship sails home from Q to P at 32km/h, find how long it will take in hours and minutes.

6. The model of a car is made to a scale 1:80. Copy and complete the table below, which refers to the car and its model.

	Length	Roof Area
Car	. . . . . m	$1m^2$
Model	3.5cm	. . . . . . cm^2

7. Give the back-bearings in each of the examples below:

(a) A is on a bearing 030° from B, so B is on a bearing . . . . . . from A.

(b) P is on a bearing 245° from Q, so Q is on a bearing . . . . . . from P.

8. The scale on a map is given as 1cm = 12km.

(a) Write this in the form 1:n.

(b) Two places on the map are 90mm apart; give the true distance between them in metres.

(c) What is the area on the map of a lake which is $720km^2$ in area?

8

EVERYDAY ARITHMETIC 1
TIME 1 HOUR

£5500

1. (a) A firm spends £76 000 on salaries; there are 3 directors earning £12 500 each and the remainder is shared equally between 7 employees.

   37 500   *38 500*

   How much does each employee earn?

   (b) One of the directors shares his job with his wife; he works 3 days a week and she does 2 days. How much does each collect from the £12 500 salary?   *7500   5000*

2. A shop is offering calculators at a reduction of 20%.

   (a) If the model on the left was marked £8.95 before the offer, what does it now cost?   *£7.16*   *80/100*

   (b) The shop originally stocked 1200 of this model which were puchased directly from the factory at £7.50 each. How much did it spend on this outlay?   *£9000*

   (c) It sold ⅔ of its stock at the pre-sale price; calculate how much it made on these.   *6884*

   (d) It sold the remainder at the sale price; how much did it take in altogether for the 1200 calculators?

   (e) State whether it made an overall profit or loss and express this as a percentage.

3. (a) Assuming that 660,000 candidates took Maths in the GCSE 1988, and 21.7% of these got a Grade C; how many candidates got a Grade C?

   (b) A school found that the number of candidates entering the 6th form in 1988 increased from 108 (1987) to 120. Express this increase as a percentage.

   (c) 15% of the candidates failed an examination; 5100 passed. How many candidates failed?

4. Mr. and Mrs. Patel have £2000 to invest; they examine two different Building Society schemes to see which offers the best deal over a year.

HAPPY HOUSE Building Society		EVEREST Building Society	
Deposit Rates		Deposit Rates	
£1 – £500	7½%	£1 – £2000	7.8%
£501 – £2000	8%	£2001 – £3000	8%

(a) How much interest would they earn on £500 for a year with Happy House?

(b) How much interest would £2000 earn with Happy House over a year?

(c) With which Society would they be better off over a year and why?

5. The dials below measure the units of electricity used by a consumer:

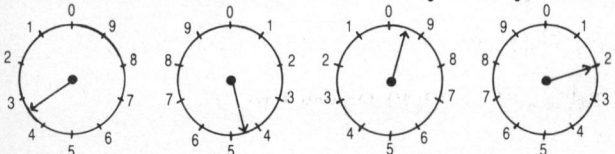

(i) (a) Write down the reading.

(b) At a subsequent reading it was found that 88 extra units were used; *sketch* 4 dials to illustrate the new reading.

(ii) A simplified version of an old bill is shown on the next page; what figures should be placed in boxes (c) to (f)?

OLD READING	1286 UNITS
NEW READING	1402 UNITS

NUMBER OF UNITS USED	*116*	(c)
COST OF UNITS @ 1.25p A UNIT	£ *1.45*	(d)
VAT @ 15%	£ .	(e)
TOTAL BILL	£ .	(f)

6. A train leaves A at 1740 and arrives at B at 2020; the distance from A to B is 256 miles.

   (a) State how long the train takes.

   (b) Calculate the average speed in miles per hour.

7. Taking £1 = 12.40 Danish Kroner

   (a) Convert 3000 Danish Kroner to sterling, giving your answer to the nearest £.

   (b) A tourist uses as a rough guide: _____ pence to the Kroner. What single figure is omitted?

8. A special paint is made by mixing 200g of Red with every 250g of Blue;

   (a) How much Red paint will there be in a 27kg drum of the mixture?

   (b) How much Red would have to be mixed with 11kg of Blue to keep the mixture consistent?

EVERYDAY ARITHMETIC 2

TIME 1 HOUR

1. (a) Copy and complete the following table for converting litres to pints:

Litres	1	2	3	4
Pints	1 ¾			

(b) How many ml are there in one pint?

2. Cooking instructions for chicken:

```
COOK FOR 25 MINS A LB
+
25 MINS OVER
```

(a) How long, in hours and minutes, does it take to cook a chicken weighing 5lbs?

(b) Mrs. Muldoon cooked her chicken for 1 hr 52½ mins; what did it weigh?

(c) The Shaw family want to eat at 2.30 p.m; at what time should an 8lb chicken be put in the oven?

3.

THE POUND	Bank Buys	Bank Sells
Australia $	2.14	2.02
Austria Sch	23.80	22.40
Belgium Fr	71.15	67.55
Canada $	2.23	2.11
Denmark Kr	13.08	12.43
Finland Mkk	7.88	7.48
France Fr	11.49	10.89
Germany Dm	3.38	3.20
Greece Dr	297	271
Hong Kong $	14.40	13.65
Ireland Pt	1.26	1.20
Italy Lira	2490	2360
Japan Yen	240	224
Netherlands Gld	3.815	3.615
Norway Kr	12.35	11.70
Portugal Esc	277	262
South Africa Rd	4.90	4.40
Spain Pta	211	200
Sweden Kr	11.60	10.98
Switzerland Fr	2.865	2.715
Turkey Lira	3600	3000
USA $	1.874	1.774
Yugoslavia Dnr	11300	9100

The chart shows the rate of exchange for the £ on a particular day. (*For example:* if you need Greek drachmas the Bank *sells* you 271 dr for £1; if you wanted to change Greek drachmas to sterling, the Bank would *buy* them from you at 297 dr to the £.)

(a) Convert £250 into Canadian dollars (assume you are a tourist going to Canada).

(b) A Bureau de Change charges 1% commission on all foreign currency transactions; calculate to the nearest penny how much a tourist from the USA would collect on changing $500 to sterling.

(c) Use the 'Bank Buys' to find how many Turkish lira are equivalent to a Japanese yen.

4.

Accommodation and Board Arrangements	MEDITERANEE Half Board
Accommodation Code	GKN
Flights Available	All Kefalonia Flts
Prices Based On	PB WC BAL
No. of Nights	7 14

		Adult	Child	Adult	Child
DEPARTURES ON OR BETWEEN	1 May – 8 May	273	191	408	273
	9 May – 23 May	292	200	428	282
	24 May – 30 May	322	260	457	346
	24 May – 30 May	322	260	457	346
	31 May – 15 Jun	308	235	454	322
	16 Jun – 29 Jun	319	241	466	327
	30 Jun – 12 Jul	332	250	479	335
	13 Jul – 19 Jul	349	262	500	349
	20 Jul – 26 Jul	364	292	518	388
	27 Jul – 2 Aug	363	291	511	382
	3 Aug – 23 Aug	358	286	506	379
	24 Aug – 6 Sep	365	272	500	350
	7 Sep – 20 Sep	340	230	478	333
	21 Sep – 2 Oct	321	211	461	323
	3 Oct – 17 Oct	286	196	423	297
	18 Oct – 24 Oct	295	205	–	–

Supplements per person per night	SV 50p; single room with bath & wc £3.90.
Reductions per person per night	3rd adult only £2.90.

The excerpt from a Thomson Holiday brochure gives details of holidays on a Greek island. You are advised to look over it carefully before answering the questions.

(a) How much would 2 adults pay for 7 nights in the hotel Mediteranee starting from the 23rd July?

(b) Mr. and Mrs. Jones took their 7 year old child on holiday for the first fortnight of the season; how much did they pay altogether?

(c) Miss Meikle went for a week's holiday and had a single room with bath and wc; she paid £391.30. When did she take her holiday?

5.

Amount of Loan	£100	£500	£1,000	£2,000	£5,000
Repayment Term: 12 Months					
Monthly payments	9.25	46.27	92.54	185.08	462.69
Total Payable	111.00	555.24	1110.48	2220.96	5552.28

The table above gives details of various loans and their repayments.

(a) Use the table to calculate how much you would pay monthly on a loan of £1500.

(b) State to the nearest whole number the annual rate percentage charged on a loan.

(c) Mr. and Mrs. Patel repaid £1332.48 over a year; use the table to deduce the size of their loan.

6. The table below shows the number of calories in various vegetables.

Vegetables	Imperial Quantity	Imperial Calories	Metric Quantity	Metric Calories
Asparagus, boiled	4oz	21	100g	18
Aubergines, raw	4oz	16	100g	14
Avocado	½	125		
Baked beans in tomato sauce	4oz	75	100g	64
Beans, broad, cooked	4oz	54	100g	48
Beans, green, cooked	4oz	22	100g	19
Beans, haricot, dried	1oz	77	25g	68
Beansprouts, canned	4oz	10	100g	9
Beetroot, boiled	4oz	50	100g	44
Broccoli, boiled	4oz	20	100g	18
Brussels sprouts, boiled	4oz	20	100g	18
Cabbage and Cauliflower, boiled	4oz	10	100g	9

(a) How many calories would you expect in a pound (lb) of boiled asparagus?

(b) How many calories would 125g of boiled broccoli contain?

(c) The table suggests that 1oz = 25g, but this is only an estimate. Is 1oz more or less than 25g?

7.

OUTWARD JOURNEY			RETURN JOURNEY		
Depart	Arrive	Service Number(s)	Depart	Arrive	Service Number(s)
MANSFIELD, Rosemary St Bus Station					
0900	1220	**450**	0125	0540	**325**
then every two hours until			0640	1005	**450**
1500	1820	**450**	then every two hours until		
1800	2120	**450**	1440	1805	**450**
1930	2250	**450**	1740	2105	**450**
2345	0405	**325**			

The timetable shows the times of coaches *from* Victoria, London (OUTWARD) to Mansfield, and the times of coaches *from* Mansfield (RETURN) to Victoria, in the course of a normal day.

(a) How long does the coach take to go from Victoria to Mansfield?

(b) How long does the last coach take to go from Mansfield to Victoria?

(c) How many coaches a day go from Victoria to Mansfield?

(d) How many coaches coming towards Victoria will the 0900 from Victoria meet?

8. Using 8km = 5 miles, convert

(a) 20km to miles

(b) 70 miles to km

(c) A car travels 35 miles on a gallon of petrol, how many gallons are required for a journey of 420 miles?

(d) A French tourist knows his car goes 12km on a litre; taking 1 gallon = 4½ litres calculate how many miles his car travels on a gallon of petrol.

GEOMETRY

TIME 1 HOUR

1. Calculate the size of each angle marked with a letter:

(a)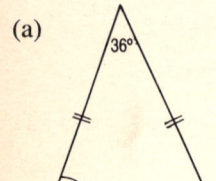

(b)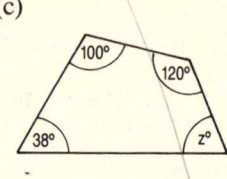

(c)

2. (a) The diagram shows the net of a cube; sketch 3 more nets, each different from the next.

(b) A certain solid has 8 edges and 5 vertices and an undisclosed number of faces; make a sketch of the net of the solid.

3.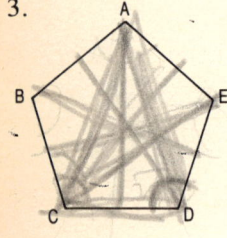

ABCDE is a regular pentagon.

(a) State the number of lines of symmetry it has.

(b) Calculate the size of angle D.

(d) Calculate the size of angle CAD.

4.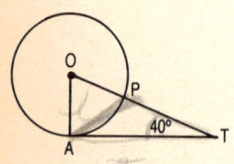

In the diagram TA is a tangent to the circle centre O. Angle T is 40°.

(a) Calculate the size of angle O.

(b) Sketch the diagram and join AP; calculate the size of angle APT.

5.   A company makes two sizes of similar toy sets for children; the triangular pieces below represent corresponding pieces, one from each set. Equal angles are marked.

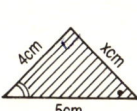 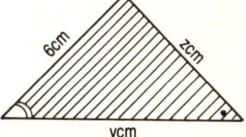

(a) Calculate the length of the side marked y.

(b) The perimeter of the smaller piece is 12cm; use this to calculate the value of x and z.

(c) Is the area of the larger piece more or less than twice the area of the smaller piece?

6.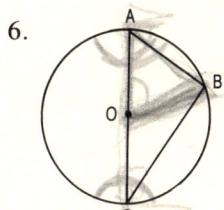

In the diagram AOC is the diameter of the circle. Sketch the diagram in your book.

(a) If angle A is 43°, calculate angle C.

(b) Join BO; what is the size of angle BOA?

7.   (a) Sketch two congruent triangles, putting in only the *minimum* information to illustrate that they are congruent.

(b) Give examples from everyday usage of a

(i) cylinder   (ii) sphere   (iii) cuboid

(c) Draw accurately an angle of 132°; state what type of angle it is.

8.   (a) 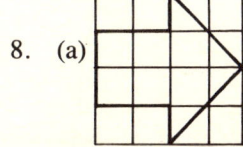   Copy the arrow shape on to a piece of squared paper; draw at least 5 more of the shape to show how it tessellates.

(b) Name one regular polygon (more than 4 sides) that

(i) tessellates   (ii) does not tessellate.

NUMBERS AND NUMBER PATTERNS
TIME 1 HOUR

1. (a) State which of the following are prime numbers:
   21, 23, 25, 27, 29.

   (b) From the set of numbers {7, 73, 169, 219} write down

   (i) a square number

   (ii) two numbers, one of which is a multiple of the other.

2. For the numbers 1.2×10^7, 9.6×10^6 and 5×10^8,

   (a) write the numbers in ascending order

   (b) add the smallest and the largest numbers, giving your answer in standard form

   (c) square the middle number, giving your answer in standard form.

3. Two joggers complete a circuit of their local park in 12 minutes and 15 minutes respectively.

   (a) If they start together at the main gate, how many laps of the park will each complete before they meet again at the main gate? How much time will have elapsed by then?

   (b) If they are joined at the main gate by a third jogger who can complete the circuit in 8 minutes, when would all three joggers meet again at the gate?

4. The pattern below is made of matches and is to be extended to the right:

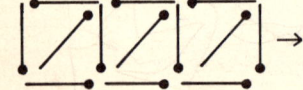

   (a) How many matches are needed to make 5 squares in the design?

   (b) If a design has 73 matches, calculate how many squares are in it.

   (c) In one pattern the number of diagonal matches added to the number of vertical matches was 37; how many horizontal matches were in the pattern?

5. Write down the missing numbers in each of the following number sequences:

   (a) 1, 4, 9, 16, 25, ____, ____

   (b) ____, ____, 17, 32, 48, 65, ____

   (c) 99, 88, 78, 69, ____, ____

   (d) 1¼, 2, 2¾, 3½, ____, ____

6. (a) The attendance at a football match was given as 26 000 (to the nearest 000). State the maximum and minimum number of spectators.

   (b) Write the number 0.0032 in standard form.

7. The patio outside Mr. Browne's house is rectangular: 168cm × 180cm; he wants to tile it using square tiles only. The factory he deals with makes square tiles with edges in whole numbers of centimetres.

   (a) If Mr. Browne decides to use 4cm square tiles, how many will he require?

   (b) What is the size of the largest possible tiles he can use?

   (c) Mr. Browne's neighbour has paved her patio with 15cm square tiles; this was the largest possible type of tile she could have used. Her patio was 120cm long; give a reasonable estimate of what its width could be.

8. (a) Arrange the following in ascending order:

   $^7/9$, $^3/4$, $^{77}/100$.

   (b) Using each of the figures 2, 3, 5, 9 once only write down a 4-digit number that is a multiple of 25.

   (c) Write down a number less than 20 that has 5 factors.

PROBABILITY AND STATISTICS
TIME 1 HOUR

1. The marks for a French spelling test set to a First Year class were:

    1, 3, 8, 7, 5, 4, 4, 5, 5, 6
    9, 10, 5, 6, 3, 5, 7, 2, 5, 9

    (a) Make a frequency table for the marks.

    (b) State the modal mark.

    (c) Calculate the mean mark.

    (d) If the pass mark was 6, state what percentage of the group failed.

2. John holds 4 cards, the Jack, Queen, King and Ace; Mina has a die with faces numbered from 1 to 6. If John plays a card and Mina throws the die, what are the chances that

    (a) the Jack and 3 will show?

    (b) the Queen and a prime number will show?

3. WELL-WEAR SHOES sent a representative to a school to make a random sample of the shoe sizes of a group of children. The results are shown in the bar-chart.

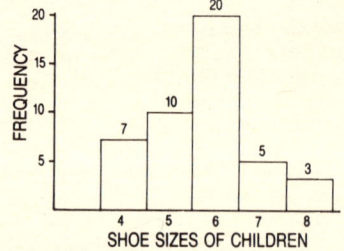

    (a) Of the three statistical measures mode, mean and median, which would be of most interest to WELL-WEAR?

    (b) If the information were shown on a pie chart, what angles would represent each size? Copy and complete the table

SIZE	4	5	6	7	8
ANGLE					

    (c) If a pupil were chosen at random, what was the probability she/he was wearing size 7 or size 8?

4.

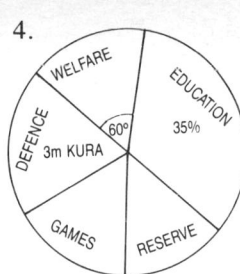

(NOT DRAWN TO SCALE)

In Kuraland where the unit of currency is the Kura, the Budget is spread over 5 areas as illustrated by the pie chart.

(a) What is the size of the angle representing Education?

(b) If the angle for Defence is 72°, state the size of the Budget.

(c) How many Kura are spent on Welfare?

(d) If Games and Reserve have the same allocation, state the size of the angle representing each.

(e) How much is held in Reserve?

5. In a survey of favourite colours, Class 3B replied as below:

Red	Blue	Green	Others
4	5	18	3

Choose a radius of your own and draw an accurate pie chart to show the preferences; label it appropriately.

6. The marks in a test were 4, 6, 9, 1, 5, 8, 3.

(a) State the median mark.

(b) If 2 pupils were chosen at random, what are the chances their marks exceeded the median mark?

7. The mean of 5 numbers is 12.

(a) Write down the total of the 5 numbers.

(b) What additional number must be included in the set to increase the mean by 1?

8.

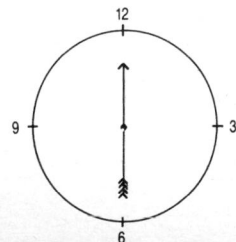

When spun, the arrow in the diagram will always stop at one of the numbers 3, 6, 9, 12. The chances of it stopping at a particular figure are given by the table:

3	6	9	12
$1/8$	$1/4$	$5/16$	x

(a) Calculate the value of x.

(b) During an experiment the arrow stopped 160 times at 9; how many times was it spun?

21

PYTHAGORAS AND TRIGONOMETRY
TIME 1 HOUR

1.

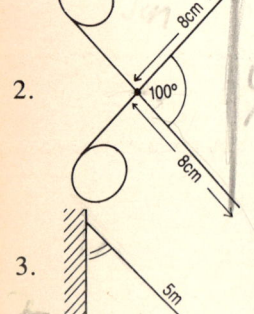

From the triangle ABC write down

(a) the length of AC

(b) (i) Sin A

    (ii) Tan B (iii) the size of angle A.

2.

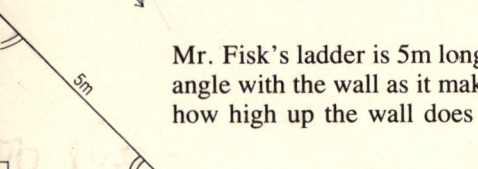

The diagram shows a scissors opened to an angle of 100°; the insides of the blades are straight lines of 8cm each. How far apart are the tips of the blades?

3.

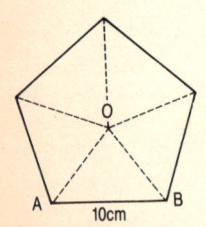

Mr. Fisk's ladder is 5m long; it makes the same angle with the wall as it makes with the ground; how high up the wall does the ladder reach?

4. To find the area of a regular pentagon (side 10cm) use the following steps:

(a) Sketch triangle AOB in your book, filling in the size of angle O, and the length of AB.

(b) Draw a perpendicular from O to AB, letting it meet AB at D. Mark in the size of angle OAD.

(c) Calculate the length of OD.

(d) Calculate the area of triangle OAB and multiply it by 5.

5. Calculate the value of x, y and z in the diagrams below:

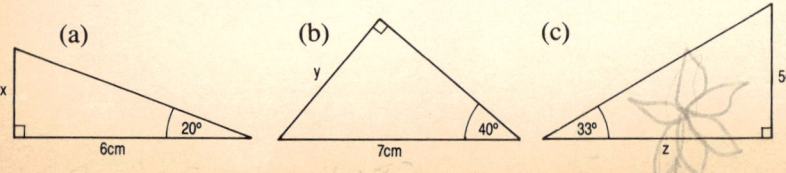

(a)

(b)

(c)

6. In a practical Mathematics lesson, some pupils were taken to the top of a school building, 10m above the level of the grounds.

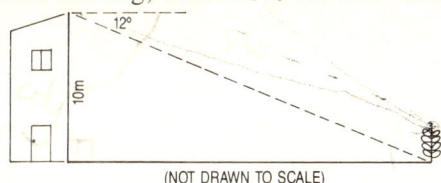

(NOT DRAWN TO SCALE)

They found that the angle of depression of the base of a shrub at the end of the playground was 12°.

(a) How far is it from the base of the building to the base of the shrub?

(b) Another group of pupils took the angle of elevation of the top of the building from the *top* of the shrub and found it to be 11°.

Use this information to calculate the height of the shrub to the nearest cm.

7. Calculate x if:

(a) Cos 2 x = 0.7.

(b) Sin ½x = 0.3.

8. A ship sails on a bearing 040° from a port P to a port, S, 60km away.

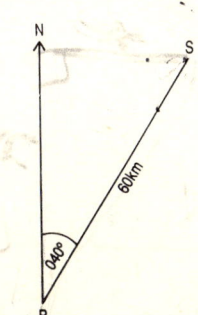

(a) How far North of its starting point is it now?

(b) How far East of its starting is it?

(c) On what bearing should it now sail if it must return to P?

(d) The Captain decides to sail due West for 20km and then go straight to P; calculate how much extra he will sail, giving your answer to the nearest 0.1km.

TRANSFORMATIONS
TIME 1 HOUR

1. (a) Plot the co-ordinates (2, 0), (0, 3) and (0, 0), join them to make a triangle.

   (b) Using the origin as the centre draw an enlargement scale factor 2.

   (c) State the ratio of the area of the smaller triangle to that of the larger.

2.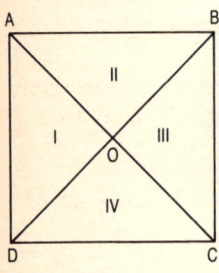

   ABCD is a square with its diagonals intersecting at O. State which transformation will map:

   (a) △ I → △ III, with A →C and D →B.

   (b) △ II → △ IV, with A →D and B →C.

   (c) △ III → △ IV, with B →D, O and C unchanged.

   (d) △ ADC → △ BCD, with A →D, C →B and D →C.

3. (a) On squared paper draw the lines x = 3 and y = 4.

   (b) Plot and join the points A(−2, 3), B(1, −2), C(−3, −3).

   (c) Reflect the triangle in the line x = 3; label it A′B′C′.

   (d) Reflect the triangle ABC in the line y = 4; label it A″B″C″.

   (e) What transformation maps A′B′C′ onto A″B″C″?

4. The points A, B, and C are three vertices of a parallelogram ABCD; A(1, 5), B(3, 6) and C(2, −1)

   (a) Plot these co-ordinates on a piece of squared paper, and complete the parallelogram.

(b) What are the co-ordinates of D?

(c) What transformation maps the line AB onto the line DC?

(d) What transformation maps the line BC onto the line AD?

5. (a) Plot and join in order (2, 3), (2, 4), (5, 4), (5, 1), (4, 1), (4, 3) and (2, 3); label it A.

(b) Rotate the shape 90° anticlockwise about (0, 0); label it B.

(c) Rotate A 270° about (0, 0) and label it C.

(d) What transformation maps B onto C?

6.

T· | T'· | T"·

ℓ_1 ℓ_2

The diagram shows an object T being reflected in two parallel lines — first in ℓ_1 then in ℓ_2.

(a) Draw accurately in your book a diagram with ℓ_1 and ℓ_2 6cm apart and T 2cm from ℓ_1. Measure the distance TT".

(b) Suppose T was 9cm from ℓ_1; what then would TT" measure?

MIXED EXAMPLES 1
TIME 1½ HOURS

1. (a) Solve the equation $1 - 3x = 10$.

   (b) Rearrange the formula $a = 5 - 3x$, to make x the subject.

2. Copy and complete the table of values for $xy = 12$.

x	−12	−6	−4	−3	−2	−1	1	2	3	4	6	12
y	−1						12					

Choose axes and scales of your own and draw the graph of the relationship. Use your graph to estimate the $\sqrt{12}$.

3. 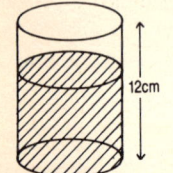 (a) What is the perimeter of the semicircle of radius 3cm?

   (b) What is the diameter of a semicircle with perimeter 100cm? (Hint: enlargement.)

4. The can of drink is ¾ full; how many cc of drink have been drunk?

5. Two tracks in a jungle run at 50° to each other; make an accurate scale drawing of the two tracks using a scale of your own, which must be stated.

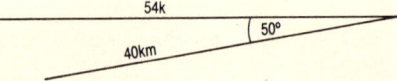

A helicopter 'shadows' two groups of tourists — one on each track. The flight-path keeps an equal distance from each track. Draw a line in your diagram to show the flight-path.

6. The scale on a map is 1:40 000.

   (a) How far apart in km are two places which are 5cm apart on the map?

   (b) A grid square on the map is 2¼ cm²; what area in m² does it represent?

7.

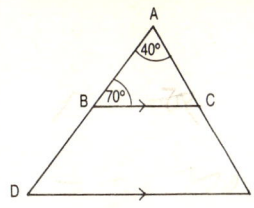

(a) What type of triangle is ABC?

(b) What is the size of angle E?

(c) If B and C are the midpoints of the sides, state the ratio of the area of the triangle ABC to the area of the trapezium BCED.

8.

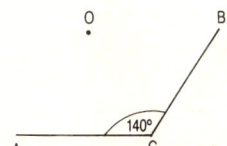

The diagram shows one corner of a regular polygon, 'centre' O;

(a) State how many sides the polygon has.

(b) What is the size of angle CAB?

9. A shopkeeper made £8 profit on a camera, which was a profit of 10%. How much did she sell the camera for?

10. A shopkeeper bought 50 radio-alarm clocks from a dealer at £21.50 each; he marked a price on each to enable him to make 20% profit.

(a) What was the marked price?

He managed to sell 80% of his stock at the marked price;

(b) How much did he take on this part of his stock?

He had to sell the remainder at 10% off the marked price;

(c) Calculate how much he had to sell each clock for and the total takings at that price.

(d) Calculate his overall profit/loss as a percentage.

11. 8 books cost £17; how much do 6 of the books cost?

12. £720 was divided in the ratio 1:x; the smaller portion was £144. Find x.

13. $a = 1.2 \times 10^4$ and $b = 1.6 \times 10^{-3}$; express in standard form

(a) ab   (b) a/b   and   (c) $\sqrt{b}$.

14. There are 3 people in a group; each sends a greetings card to every other member in the group; how many greeting cards are required?

Suppose there were 4 people in the group; how many cards would then be needed? Can you state a general rule for the number of cards needed in a group of n people?

15. The probability of winning a certain game is 2/5, and of drawing 1/10;

    (a) What are the chances of losing a game?

    (b) In a series of games Joan won 16 times; how many games were likely to have been played?

16. A group of children displayed their preferences for TV programmes by a pie chart. The table shows the angles required:

NEIGHBOURS	EASTENDERS	COR'. STREET	OTHERS
120°	75°	105°	?°

    (a) What angle should replace the ?° ?

    (b) If 35 children watched Eastenders, how many were interviewed altogether?

17. If Sin x = 0.3, find   (a) x   and   (b) Cos x.

18. 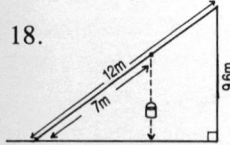 A painter standing 7m along a ladder drops a tin of paint; how far from the bottom of the wall will the tin hit the ground?

19. Draw the triangle ABC where A is (1, −3), B(2, 2) and C(−4, −1). Rotate the triangle 270° about the origin.

20. If the triangle in question 19 is translated using the vector

$$\begin{pmatrix} 2 \\ -1 \end{pmatrix}$$, state its new co-ordinates.

MIXED EXAMPLES 2
TIME 1½ HOURS

1. (a) Simplify $4(3x - 5y) - 3(4x - 2y)$.

   (b) Factorise $ax^2 + x$.

2. Choose values of x and y to draw the straight line $2y = 12 - 3x$. State the gradient of the line.

3. 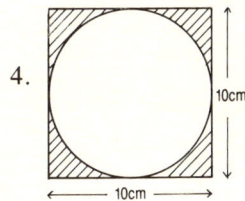 The volume of this cylinder is 40cm³; calculate its length, if the diameter is 7cm.

4. Calculate the area of the shaded part of the diagram, giving your answer to 1 decimal place.

5. 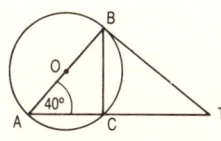 Draw accurately the parallelogram on the left; divide it into 2 congruent trapezia.

6. What is the back-bearing for each of the following:

   (a) 070°,   (b) 130°   and   (c) 320°?

7. TB is a tangent to the circle with diameter AOB.

   (a) Calculate the size of angle T.

   (b) State a geometrical property that the three triangles in the diagram have in common.

8. Sketch two triangles which are similar but not congruent; you must give measurements to the three sides of each triangle.

9. An article was marked £8.25 before a sale, during which it was offered at a discount of 12%. What would a customer have paid for it during the sale?

10. Mr. and Mrs. Lewin run a stall for 48 hours a week; they divide the time in the ratio 3½:2½. How many hours a week does each of them work?

11. A 75cl bottle of wine costs £2.70; how much should 100 cl of the same wine cost?

12. Taking 1 inch = 2½cm, convert

   (a) 18 inches to cm

   (b) 1m to inches.

13. Write down the missing numbers in these number sequences:

   (a) 5, 6, 8, 11, 15, _____, _____.

   (b) _____, ⅝, 1¼, 2½, _____.

14. The pattern of matches can be extended to the right; the head of a match always faces either 'East' or 'South'.

   (a) How many matches will there be in a pattern with 15 'South' matches?

   (b) In an extended pattern there are 99 matches; how many East matches are there in the pattern?

15. (a) If a die is thrown 1800 times how many times would you expect a prime number to turn up?

   (b) If 2 dice are thrown and the results added together, what are the chances the score is an even number?

16. Favourite soup for Class 1C:

MINESTRONE	TOMATO	DALL	PEA
7	5	8	10

Draw an accurate pie chart to show 1C's soup preferences.

17. *Sketch* a regular 9-sided polygon, centre O, side 12cm. Calculate the area of the polygon by following the steps:

(i)   Sketch triangle OAB (where A, B are two adjacent vertices of the polygon); mark in the size of angle O and the length AB.

(ii)   Draw a perpendicular from O to AB, letting it meet AB at D; calculate the size of angle OAD.

(iii) Calculate the length of OD.

(iv)   Calculate the area of triangle OAB and multiply it by 9.

18.
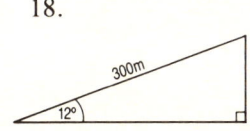

A road slopes up 12°, to the horizontal; how much will you rise by walking 300m along the road?

19. The co-ordinates (0, 4), (8, 0) and (12, 12) are the vertices of a triangle which has been enlarged by scale factor 4, centre (0, 0). State the co-ordinates of the original triangle.

20. The co-ordinates of two triangles are given below:

△ I: (−2, 3), (−3, −4) and (0, 0) and
△ II: (3, 2), (−4, 3) and (0, 0).

State the transformation that maps triangle I onto triangle II.

PART II

ANSWERS TO PRACTICE PAPERS CONTAINED IN PART I

ANSWERS

ALGEBRA AND GRAPHS

1. (a) 4½   (b) 12½

2. (a) (i) (2, 0) (ii) (0, 6)   (b) -3

3. (a) $C = \frac{1}{2}n + 5$ (b) £12.50   (c) 6
    (d) $n = 2c - 10$ (e.g.)

4. (a) $12y (x + 2z)$ (b) $12ab - 20ac$ (c) $-3x + 2y$

5. (a) ± 15 (b) -2

6. (a) 12, 6¾, 3, ¾ etc.   (c) 6   (d) ± 1.73

7. (a) 7, 11   (b) (0, 3)   (c) 2

8. (3, 3)

AREA AND VOLUME

1. (a) 8m   (b) $24m^2$

2. (a) $462cm^2$ (b) 4.95

3. (a) 600   (b) £351

4. (a) 47   (b) 193.2

5. (a) 50m   (b) 124

6. (a) 14.4   (b) 21.78

7. 6

8. (a) 8   (b) 8   (c) 14cm

CONSTRUCTIONS (including bearings and scales)
Answers may differ slightly.

1. (b) 4.3

2. (a) 100 miles approx.   (b) 030°   (c) 240°

3. (b) 270°

4. (a) 625km   (b) 46mm

5. (c) 18hrs 45mins

6. 2.8, $1^9/_{16}$

7. (a) 210°   (b) 65°

8. (a) 1:1,200,000   (b) 108000   (c) 5cm^2

EVERYDAY ARITHMETIC 1

1. (a) £5500   (b) £7500, £5000

2. (a) £7.16   (b) £9000   (c) £7160   (d) £10024   (e) 11.4% profit.

3. (a) 143220   (b) 11.1%   (c) 900

4. (a) £37.50   (b) £157.50   (c) HH by : £1.50

5. (a) 3492   (c) 116   (d) £1.45
    (e) 22p   (f) £1.67

6. (a) 2 hrs 40 mins   (b) 96 m.p.h.

7. (a) £242   (b) 8

8. (a) 12kg   (b) 8.8kg

EVERYDAY ARITHMETIC 2

1. (a) 3½, 5¼, 7   (b) 571

2. (a) 2 hrs 30 mins   (b) 3½   (c) 10.45am

3. (a) 527.50   (b) £264.14   (c) 15

4. (a) £728   (b) £1089
    (c) Between 20th − 26th July

5. (a) £138.81   (b) 11%   (c) £1200

6. (a) 84   (b) 22½   (c) more

7. (a) 3 hrs 20 mins   (b) 3 hrs 25 mins
    (c) 7   (d) 3

8. (a) 12½   (b) 112   (c) 12   (d) 33¾

GEOMETRY

1. (a) 72°   (b) 144°   (c) 102°

2. Various

3. (a) 5   (b) 108°   (c) 36°

4. (a) 50°   (b) 115°

5. (a) 7½   (b) 3, 4½   (c) more

6. (a) 47°   (b) 94°

7. (c) obtuse

8. (b) (i) Hexagon   (ii) Pentagon (for example)

NUMBERS AND NUMBER PATTERNS

1. (a) 23, 29   (b) (i) 169   (ii) 219:73

2. (a) 9.6x, 1.2x, 5x   (b) 5.096×10^8 (c) 1.44×10^{14}

3. (a) 5, 4, 60 mins   (b) 2 hrs later

4. (a) 21   (b) 18   (c) $18 \times 2 = 36$

5. (a) 36, 49   (b) -10, 3, 83   (c) 61, 54   (d) 4¼, 5

6. (a) 25500 and 26499   (b) 3.2×10^{-3}

7. (a) 1890   (b) 12cm side   (c) 105 or 135 (e.g.)

8. (a) $^3/_4$, $^{77}/_{100}$, $^7/_9$ (b) 9325 or 3925   (c) 16

PROBABILITY AND STATISTICS

1. (b) 5   (c) 5.45   (d) 60%

2. (a) $^1/_{24}$ (b) $^1/_8$

3. (a) Mode   (b) 56, 80, 160, 40, 24 degrees
    (c) 8/45

4. (a) 126°   (b) 15 million Kura (c) 2½m
    (d) 51°   (e) 2⅛m

5. Angles 48, 60, 216, 36 degrees

6. (a) 5   (b) $^1/_7$

7. (a) 60   (b) 18

8. (a) $^5/_{16}$ (b) 512

PYTHAGORAS AND TRIGONOMETRY

1. (a) 6   (b) (i) 0.8   (ii) 1¾   (iii) 53.1°

2. (a) 12.3cm

3. 3.54m

4. (a) 72°   (b) 54°   (c) 6.88   (d) 172cm^2

5. (a) 2.18   (b) 4.50   (c) 7.70

6. (a) 47.0m   (b) 86cm

7. (a) 22.8°   (b) 34.9°

8. (a) 46.0km   (b) 38.6   (c) 220°   (d) 9.6km

TRANSFORMATIONS

1. (c) 1:4

2. (a) 180° rotation about O
    (b) Reflection in line parallel to AB through O.
    (c) Reflection in OC   (d) Rotation 90° a/clockwise about O.

3. (e) Rotation 180° about (3, 4)

4. (b) (0, −2)

    (c) translation $\begin{pmatrix} -1 \\ -7 \end{pmatrix}$ (d) translation $\begin{pmatrix} -2 \\ -1 \end{pmatrix}$

5. (d) Rotation 180° about (0, 0)

6. (a) 12cm   (b) 12cm

MIXED EXAMPLES 1

1. (a) -3 (b) $\dfrac{5-a}{3}$

2. ± 3.5 approx

3. (a) 15.4   (b) 38.9

4. 151

5. Angle bisector

6. (a) 2km   (b) 360000

7. (a) Isosceles   (b) 70°   (c) 1:3

8. (a) 9   (b) 20°

9. £88

10. (a) £25.80   (b) £1032   (c) £232.20   (d) 17.6% gain

11. £12.75

12. 1:4

13. (a) 1.92×10^{1} (b) 7.5×10^{6} (c) 4×10^{-2}

14. (a) 6   (b) 12   (c) $n(n-1)$

15. (a) ½   (b) 40

16. (a) 60°   (b) 168

17. (a) 17.5°   (b) 0.954

18. 3m

19. Diagram

20. $(3, -4)$, $(4, 1)$, $(-2, -2)$

MIXED EXAMPLES 2

1. (a) $-14y$ (b) $x(ax + 1)$

2. $-1\frac{1}{2}$

3. 1.04

4. 21.5

5. Any line through the 'centre' from side to side.

6. (a) 250°   (b) 310°   (c) 140°

7. (a) 50°   (b) similar

8. Diagrams

9. £7.26

10. 28, 20

11. £3.60

12. (a) 45cm   (b) 40 in

13. (a) 20, 26   (b) $^{5}/_{16}$, 5

14. (a) 29   (b) 49

15. (a) 900   (b) ½

16. Angles: 84, 60, 96, 120 degrees

17. (i) 40°   (ii) 70   (iii) 16.5   (iv) 890

18. 62.4

19. (0, 1), (2, 0) and (3, 3)

20. Rotation 270° about (0, 0)

PART III

GCSE PRACTICE PAPERS

TIME ALLOWED
1 HOUR EACH PAPER

INTRODUCTION AND CONTENTS

There are ten papers, each designed to take an hour and testing a different topic (or a small group of topics).

There are also two longer papers of mixed examples for extra practice before Mock Examinations or the main examination. They also provide additional material should you wish to ''create'' further practice papers.

ALGEBRA AND GRAPHS

TIME 1 HOUR

1. Solve the equations:

   (a) $\dfrac{3x + 1}{2} = 7$,   (b) $5 - 4x = -9$.

2.

x	0	3	
y			4

   (a) Copy and complete the table of values for the relationship $2x + 3y = 12$.

   (b) Plot your values for (x, y) and join them with a straight line; state the gradient of the line.

3. CUTTALAWN offers to mow your lawn for a fixed fee of £50 plus £1 per square metre. (i.e. $C = 50 + n$ where n is the area of the lawn in m^2, and C the total cost).

BETTALAWN offers to do the job for a fixed fee of £30 plus £2 a square metre. (i.e. $C = 30 + 2n$).

   (a) Copy and complete the table below for lawns of various sizes and the respective costs of the two companies:

square m: n	5	10	15	20	25	30
Cuttalawn: £C	55					
Bettalawn: £C					80	

   (b) Label a sheet of graph paper as below and draw two graphs to show the respective costs for mowing lawns of area n m^2. Label your graphs clearly.

   (c) For what size of lawn does each company charge the same?

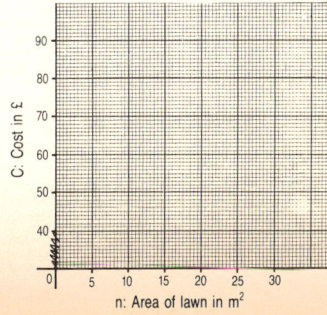

43

(d) What is the difference in costs between the companies for a lawn $12m^2$?

(e) For what sizes of lawns does one company charge £12 less than the other?

4. (a) Factorise $3x^2 + 3x$

(b) Expand $6x(x - 3y)$

(c) Simplify: $4(2x - 5y) + 3(5y - 2x)$

5. (a) Given that $z = x^2 + y^{-2}$, calculate z when $x = -2$ and $y = 2$

(b) Given that $ax + by = cz$, rewrite the formula to make y the subject.

6. For the relationship $y = -2x^2$

(a) copy and complete the table of values

x	−2	−1½	−1	0	½	1½	2
y				0			−8

(b) Using axes and scales of your own plot the co-ordinates (x, y) and join them with a smooth curve.

(c) Join the points (0, 0) and (2, −8) with a straight line; write down the gradient of this line.

(d) *Use your graph* to solve the equation $-2x^2 = -6.5$, making clear your method.

7. What is the missing line in this Algebra sum

$$+ \quad \begin{array}{ccc} 3x & - & 7y \\ \bigstar & \bigstar & \bigstar \quad ? \\ \hline -x & + & 10y \end{array}$$

8. What is the cost in £ of q items at x pence each and r items at £5 each?

AREA AND VOLUME
TIME 1 HOUR

1. The top of a coffee-table is rectangular, with a sheet of glass inset as shown. The wood is 5cm wide all around.

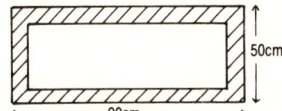

(a) Calculate the area of the glass.

(b) Calculate the area of the wood.

2. A can of Refresh has a base radius of 4cm and a height of 15cm.

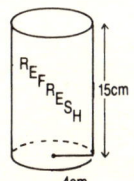

(a) Calculate, to the nearest $10cm^3$, the volume of the can.

(b) For practical purposes, what fraction of a litre does the can hold when full?

3. Mrs. Jones' patio is covered with square tiles 45cm × 45cm; the patio measures 8 tiles by 5 tiles.

(a) What is the area of 1 tile in cm^2?

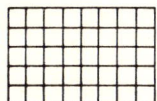

(b) What is the perimeter of the complete patio in metres?

(c) What is the area of the patio in m^2?

(d) Tiles are sold in boxes of 5 for £7.20; how much did it cost to cover the patio?

4. Soupo cubes are 1½cm along each edge and are packed into boxes measuring 12cm × 6cm × 9cm.

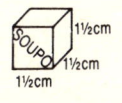

(a) What is the volume of 1 cube of Soupo? (You may give your answer in decimal form.)

(b) How many cubes will fit into a box?

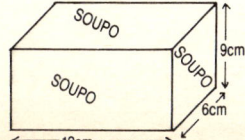

(c) The Soupo box is wrapped in silver foil; calculate the area of silver foil required, ignoring overlaps. You are advised to set your work out as follows:

FRONT: ? × ? = ... cm^2
BACK: etc.

5. The diagram shows a school racing track with semi-circular ends.

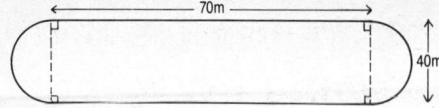

(a) How much is 1 complete lap of the track, to the nearest 5m?

(b) A pupil in training needs to run 3km in a session; how many laps of the track should she run?

(c) The space inside the track is to be planted with grass-seed; if 1 bag of grass-seeds covers 120m², how many bags are required to complete the planting?

6.

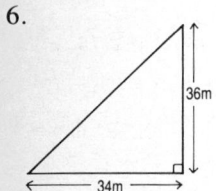

In measuring this piece of triangular ground, an error of ± 10cm was acceptable.

(a) Calculate to the nearest ½m², the maximum area that could be given for the patch of ground.

(b) Calculate in cm², the area of a drawing of the triangle on a scale of 1:100.

7.

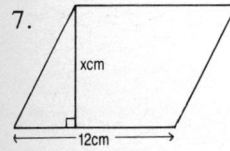

If the area of the parallelogram is ⅓ the area of a square of side 9cm, calculate x.

8.

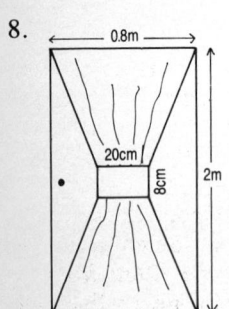

The diagram shows a glass door partly covered by a curtain; the curtain is held in shape with a 20cm × 8cm band around its centre.

Calculate the area of glass not covered by the curtain. (There are different ways to approach this question – give it some thought before you start.)

CONSTRUCTIONS
(including bearings and scales)

TIME 1 HOUR

1. (a) Construct a triangle with sides 6cm, 8cm and 10cm.

    (b) Measure the smallest angle in the triangle.

    (c) What is the length of the perpendicular from the vertex to the longest side?

2.

The diagram is taken from Thomson Holiday Summer Brochure.

(a) Use the scale to estimate the distance from Cala'n Forcat to Arenal D'en Castell.

(b) What is the bearing of Mahon from Arenal D'en Castell?

(c) What is the bearing of Cala Galdana from Arenal D'en Castell?

3. The positions of 3 lakes are shown in the diagram, with L_2 directly North of L_3.

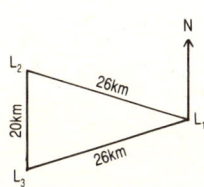

(a) Using a scale of 5mm = 1km, draw an accurate diagram of the 'triangle'.

(b) What is the bearing of L_2 from L_1?

(c) It is proposed to build an industrial complex within the 'triangle' $L_1L_2L_3$; for conservation purposes it must be more than 12km from L_1 and more than 8km from the imaginary line joining L_2 and L_3. On your diagram indicate clearly the area available for the complex.

4. *The scale on this map is 1:5000 000*

(a) How far apart, in km, are Palermo and Catania?

(b) If two towns in Sicily are 70km apart, how far apart are they on the map?

5. A plane flies on a bearing of 320° from City A to City B, a distance of 1200 miles.

(a) Using a scale of your own draw an accurate diagram to illustrate its route — state your scale.

(b) If the plane averages 500 m/h, calculate in hours and minutes how long the journey takes.

(c) On what bearing must it fly on the return journey from B to A?

6.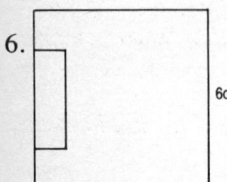

The diagram, a square of side 6cm, is a scale drawing of a classroom. The actual teacher's desk measures 150cm × 120cm. Copy and complete the table below which refers to the classroom.

	True Area	Model Area
Classroom	29.16m^2	36cm^2
Teacher's Desk	. . . . . . m^2	. . . cm^2

7. State the back-bearings for each of the following bearings:

(a) 130°   (b) 009°   (c) 300°   (d) 200°.

8. The scale on a map is 1cm = 7½km.

(a) Write this in the form 1:n.

(b) On the map two places are ¾cm apart; how far apart are they in km?

(c) A grid square on the map is 6¼cm^2; what is the length, in km, of the side of the square it represents?

48

EVERYDAY ARITHMETIC I
TIME 1 HOUR

1. Mandy accepts a job with a company on a salary of £9600 a year.

    (a) If she pays 32% in 'stoppages' out of her salary, how much does she 'clear' a month?

    (b) She shares a flat with her sister who earns £8000 a year; they agree to share the weekly rent of £110 in the ratio of their salaries. How much does each contribute towards the rent?

2. Roxy's show is being held in the Old Pavilion which seats 2,000 people; the owners of the Old Pavilion provide ⅞ of the tickets for sale and the remainder are sold through an agency.

    (a) How many tickets does the Old Pavilion get to sell, and how much will they take in if the tickets cost £17.50 each?

    (b) A fan of Roxy who could not get a ticket from the Old Pavilion went to the agency which asked £24.50 for one. What percentage 'mark-up' does this represent over the Pavilion's price?

    (c) If the agency sold all its tickets at this price, calculate how much it took for its allocation of tickets.

3. Assume 660 000 pupils took GCSE Maths in 1988 and 16% got Grade E.

    (a) Calculate how many pupils got Grade E.

    (b) Grade E could have been awarded to candidates taking either the Intermediate or Foundation Level papers; assuming that 5 times as many got Grade E through the Foundation Level papers as through the Intermediate, calculate how many candidates got Grade E (Foundation Level).

    (c) 46,200 candidates got Grade G; what percentage of the entry total was this?

4. Kevin wants to buy on Hire Purchase a stereo system which costs £700 for cash; the store requests a deposit of 20% on all Hire Purchase agreements.

    (a) Calculate the deposit he will have to pay on the stereo.

In addition to the deposit he has to pay 12 monthly instalments of £52 each.

(b) How much will he pay altogether for the stereo?

(c) Calculate, to the nearest whole number, the percentage profit the shop makes on the stereo.

5. A motorist hired a car and noted the mileage reading before her journey was

1	9	8	5	4

; at the end of the first day the reading was

2	0	2	2	4

.

(a) How many miles did she travel that day?

(b) If the car did 34 miles to a gallon of petrol, calculate how many gallons of petrol she needed, giving your answer to the nearest whole number.

(c) The next day she used 2½ gallons of petrol on a journey; give the reading for the mileage at the end of that day.

(d) The petrol for the 2 days cost £24.30; calculate to the nearest 5p the cost of a gallon of petrol.

6. An insurance company offers to insure household goods at £6.50 per £1000 of possessions.

(a) Calculate the premium payable on possessions worth £8000.

(b) Calculate the value of the possessions of a house where the premium is £156.

7. Part of a recipe to *SERVE 4 PEOPLE* is given below;

½ pt (300ml) of apple juice
16oz (454g) of fresh pineapple

\-   \-   \-   \-
\-   \-   \-   \-
\-   \-   \-   \-
\-   \-   \-   \-

Rewrite the two lines of the recipe to *SERVE 6 PEOPLE*.

8. (a) Divide £420 in the ratio 1:2:4.

(b) John has 3 times as much money as Jane; between them they have £15; how much each of them got;

(c) A sum of money was divided in the ratio 1:5:9. If the middle share was £25, what was the largest share?

EVERYDAY ARITHMETIC 2
TIME 1 HOUR

1. (a) Copy and complete the table below to convert kilograms to pounds:

K'grams	1	2	4	8
Pounds	$2^1/_5$			

(b) Use the table to convert 1lb to grams − giving your answer to the nearest gm.

2. The chart shows the distances between various places in Great Britain:

A'deen

626	Dover			
584	246	Exeter		
543	77	170	London	
322	292	291	207	York

(a) How much will a motorist cover on the route ABERDEEN − DOVER − EXETER?

(b) A travelling salesperson went from X − Y − Z; if the combined trip was 323 miles which 3 towns were involved?

(c) A driver went from Aberdeen to York at an average speed of 70 mph; find how long the journey took.

3. The graph converts £ sterling to Hong Kong dollars and vice versa.

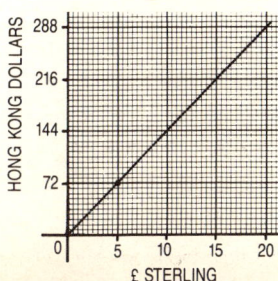

(a) Convert £3.50 to HK dollars.

(b) Convert 180 HK dollars to sterling.

(c) Use the graph to convert £35 to HK dollars, explaining your method.

4.

Accommodation and Board Arrangements	ES PORT Bed & Breakfast			
Accommodation Code	MST			
Flights Available	Tue and Sat Flights			
Prices Based On	PB WC BAL			
No. of Nights	7		14	
	Adult	Child	Adult	Child
13 Apr–30 Apr	131	79	185	113
1 May–8 May	141	89	200	128
9 May–23 May	165	105	227	145
24 May–30 May	193	144	255	189
31 May–15 Jun	179	124	250	173
16 Jun–29 Jun	188	140	279	193
30 Jun–12 Jul	207	154	301	208
13 Jul–19 Jul	229	171	330	229
20 Jul–26 Jul	240	191	353	263
27 Jul–2 Aug	238	190	345	258
3 Aug–23 Aug	238	190	339	253
24 Aug–6 Sep	246	183	326	228
7 Sep–20 Sep	225	153	305	212
21 Sep–2 Oct	215	145	292	202
3 Oct–17 Oct	190	131	258	179
18 Oct–24 Oct	213	149	–	–
Supplements per person per night	HB £3.00; FB £7.00; single room with sh & wc £2.60.			
Reductions per person per night	3rd adult only £1.20.			

(DEPARTURES ON OR BETWEEN)

The excerpt from a Thomson Holiday brochure gives details of holidays on a Spanish island. You are advised to look over it carefully before answering the questions.

(a) How much will a family of 2 parents and 3 children pay for a 14-day holiday at the resort if they depart on 15th July?

(b) A single person using a single room with shower and wc was charged £258.20 for 7 nights; during which period did she depart?

(c) The flight from Gatwick takes 2 hrs 15 mins and arrives at 0110; state the departure time.

5. The travel-graph shows a coaches trip from town A to town B, 160 miles away.

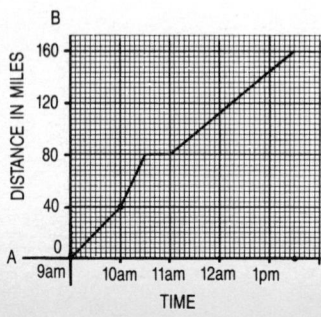

(a) What was the average speed of the coach for the part of the journey when it was moving fastest?

(b) What was the time when the coach was 60 miles from B?

(c) Another coach left B for A at 9 a.m. and travelled at a constant speed of 60 mph. Estimate from the graph the time when the two coaches met.

6. (a) If 8 pineapples cost £5, how much should 6 cost?

   (b) How many such pineapples could be bought for £12.50?

7. What figures should be placed in the empty spaces in the bill below:

	£	.	p
3 MARS BARS @ 19p each: =		.	
☐ CHOCOS @ 17p each: =	1	.	87
☐ CRISPS @ 12p each: =		.	
TOTAL : =	4	.	00

8. 4 cleaners can clean a school in 30 minutes:

   (a) How long would it take 1 cleaner?

   (b) If the Local Authority reduces the number of cleaners to 3, how much longer will each of the other cleaners now have to work in order to clean the school?

GEOMETRY

TIME 1 HOUR

1. 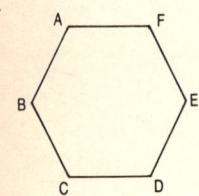 ABCDEF is a regular hexagon.

(a) Calculate the size of angle D.

(b) Calculate the size of angle BFC.

(c) How many lines of symmetry has the shape?

2. (a) State which, if any, of the arrangements below are nets of a cube.

(i) 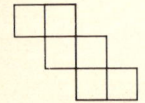   (ii) 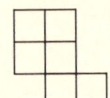   (iii)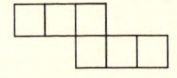

(b) The diagram shows part of the net of a closed box; sketch it in your answer book, and complete the net — showing necessary measurements.

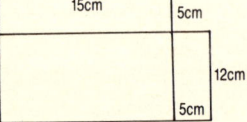

3. 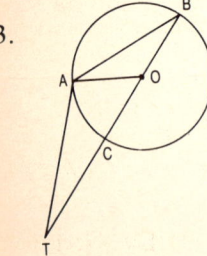 In the diagram TA is a tangent to the circle centre O. Angle B is 30°.

Calculate (a) the size of angle T

(b) the size of angle CAT.

4. 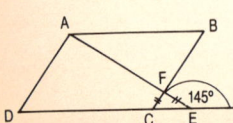 ABCD is a parallelogram; AF and DC produced meet at E, and FC = FE.

Calculate the size of (a) angle D, and

(b) angle DAE.

5.

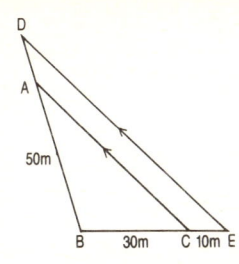

The diagram shows part of the corner of a field with two parallel electrical fences to separate different animals.

(a) Calculate the length of BD.

(b) If the wire from A to C is 75m, how long is the wire from D to E?

(c) If the area of triangle ABC is 504m², calculate the area of triangle DBE.

6. (a) Name 3 different parts of a circle (excluding centre, radius, circumference).

(b) Give examples from every day usage of

(i) cone   (ii) cube.

(c) Draw an acute angle of your own, stating its size.

7.

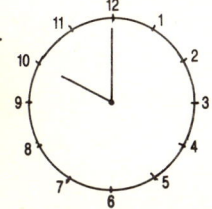

(a) What is the size of the acute angle between the hands of the clock at 10 o'clock?

(b) By how much will this angle have increased by 10.20?

8. (a) On squared paper draw a scalene triangle (no two sides the same); by drawing at least 5 more triangles, show how to tessellate with your triangle.

(b) Name a quadrilateral that has:

(i)   equal diagonals which bisect each other at angles different from 90°

(ii) rotational symmetry of order 2, and diagonals that bisect each other at 90°.

NUMBERS AND NUMBER PATTERNS
TIME 1 HOUR

1. (a) State which of the following are square numbers:

    1, 19, 64, 89, 1000.

    (b) From the set of numbers {16, 17, 217, 480} write down:

    (i) a prime number

    (ii) two numbers, one of which is a multiple of the other.

2. (a) Arrange the numbers 2×10^{-3}, 1.3×10^{-4} and 5×10^{-2} in ascending order.

    (b) Write in standard form the difference between the largest and smallest numbers in part (a).

    (c) Write $\sqrt{289}$ in standard form.

3. Two bells chime at 8 and 12 minute intervals respectively.

    (a) If they chime together at 11 a.m., when will they next chime together?

    (b) At which hour of the clock will they next chime together?

4. A class of children were told to investigate how many handshakes take place in a group when each person shakes once with everybody else.

    (a) Copy and complete the table below to show what they should have found:

Number in Group	2	3	4	5	6
Number of Handshakes	1	3			

    (b) How many people would have to be in a group for 45 handshakes to take place?

    (c) Could you have a group where 60 handshakes take place?

5. Write down the missing numbers in the following number sequences:

    (a) 25, 16, 9, 4, 1, ¼, _____, _____.

    (b) _____, _____, 3, 8, 14, 21, _____.

    (c) _____, 1½, 4½, 13½, 40½, _____.

    (d) 17, 8, 15, 11, 13, 14, _____, _____.

6. (a) The area of a circle was given as 78.5cm^2, to the nearest 0.5cm^2; complete the blank spaces in this sentence: the area could be equal to or greater than . . . . or any number up to but not including . . . .

    (b) Write the number ninety-three million in standard form.

7. A glass factory makes large sheets of glass which are then cut into squares of varying sizes – the length of the side of the square is always a whole number of cm.

From a sheet 2m × 2.4m,

    (a) how many squares of side 4cm can be cut?

    (b) how many identical squares can be cut if the edge is to be as long as possible?

    (c) suppose they cut 480 small squares from a sheet; what would be the length of the edge of each square?

8. (a) Arrange in ascending order:

    $$\left(\frac{5}{7}\right)^2 \; ; \; 50\%; \; 0.501.$$

    (b) Using each of the figures 8, 5, 1, 7 only once write down a 4-figure number which is a multiple of 25.

    (c) Write 20 as the sum of 3 prime numbers:

    _____ + _____ + _____ = 20

PROBABILITY AND STATISTICS
TIME 1 HOUR

1. Two dice are thrown together; what are the chances that:

   (a) each shows a six

   (b) the sum of the two scores is 7?

2. The bar chart below represents the age groups of a crowd attending a boxing match.

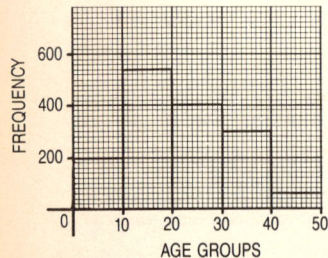

   (a) How many people between the ages of 10 and 20 attended?

   (b) How many people attended altogether?

   (c) In which age group did the median spectator occur?

   (d) There is a 1/5 chance of a spectator chosen at random coming from a particular age group; which age group is being referred to?

3. The mean average of 8 numbers is 10.

   (a) What is the sum of the numbers?

   (b) One of the numbers is now removed and the mean goes up by 1; which number has been removed?

4. BIGGA-STORE checked a consignment of various joints of beef under three categories: U(nderweight), E(xact), and O(verweight). The table shows their findings:

U	E	O
1¼%	98%	●%

   (a) What figure should replace the smudge (●)?

   (b) The chances of any joint of beef being contaminated is estimated by BIGGA-STORE to be 0.2%. In a consignment of 5000 joints, how many would they expect to be of exact weight and contaminated?

   (c) Why would a barchart be inconvenient to illustrate the information in the completed table?

5. 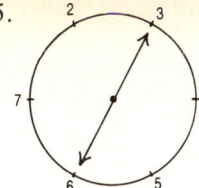 When the double headed arrow stops spinning it always points directly towards 2 numbers; what are the chances of:

(a) it pointing to 4↔7?

(b) it pointing to 2 prime numbers?

6. The table below shows the frequency of marks in a test:

Mark	1	2	3	4	5	6	7	8
Freq.	0	3	3	7	8	3	3	3

(a) What was the modal mark?

(b) What was the median mark?

(c) Calculate the mean mark.

7. 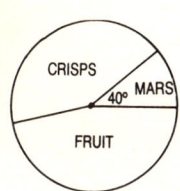 The pie chart illustrates the preferences of a group of children for Mars, Crisps or Fruit. The angle representing Mars is 40° − equal numbers said Crisp and Fruit.

(a) Using a radius of your own, draw a pie chart to show the preferences.

(b) If 12 children said Mars, how many:

    (i)   said Crisps?

    (ii) were interviewed?

8. From an ordinary pack of playing cards, what are the chances of selecting:

(a) a King if only one card is drawn out.

(b) a King or a spade if only one card is drawn out.

(c) a King followed by a King if the first card is kept out?

PYTHAGORAS AND TRIGONOMETRY
TIME 1 HOUR

1.

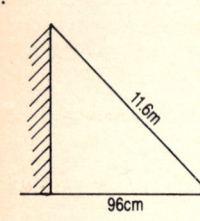

The diagram shows a ladder, 11.6m long leaning against a wall; the foot of the ladder is 96cm from the wall.

(a) How high is the top of the ladder above the ground?

(b) What angle does the ladder make with the ground?

2. Calculate x, y and z in each of the diagrams below, giving your answers to the nearest 0.1 in each case.

(a)

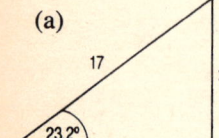

(b)

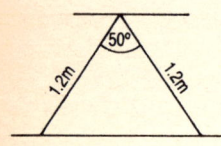

(c)

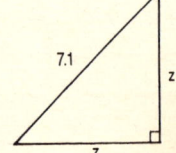

3.

The diagram shows a pair of steps open 50°.

(a) How far apart are the feet of the steps?

(b) What is the height of the top of the steps above the ground?

4.

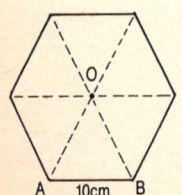

To find the area of a regular hexagon use the following steps:

(a) Sketch triangle AOB filling in the size of angle O.

(b) Draw a perpendicular from O to AB and calculate its length.

(c) Find the area of triangle AOB and multiply it by 6.

5. (a) Sin 2x = 0.6; calculate x.

(b) Cos (y − 30) = 0.5; calculate y.

6. A practical lesson in Mathematics took place in the local park; the teacher's instructions to a group of pupils were, "walk North 60m from the gate (G), then East 45m where I will meet you".

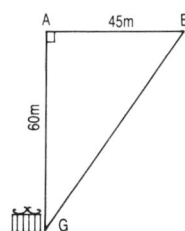

(a) The teacher walked directly from G to B; how much less did she walk than the pupils?

(b) On what bearing did the teacher walk?

(c) If the group returns directly to the gate, what bearing will they take?

7. To measure the height of a tree Gina walked 20m from its base and measured the angle of elevation, which she found to be 51°.

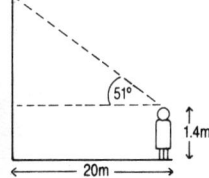

(a) Calculate the height of the tree to the nearest m.

(b) What is the angle of depression of Gina's feet from the top of the tree?

8. A road slopes up at 10° to the horizontal; how much will you rise if you walk 700m along the road?

TRANSFORMATIONS
TIME 1 HOUR

1. (a) Plot the co-ordinates $(-1, 0)$, $(0, 2)$ and $(0, -1)$, join them to make a triangle.

   (b) Using the origin as centre, draw an enlargement scale factor 2.

   (c) Write down the area of each triangle, and give the ratio of their areas in simplest form.

2. Two squares have a common side DC as shown; state which transformation maps:

   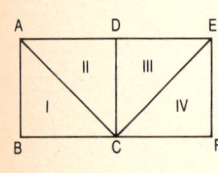

   (a) $\triangle$ I $\rightarrow$ $\triangle$ IV with A $\rightarrow$E and B $\rightarrow$F.

   (b) $\triangle$ II $\rightarrow$ $\triangle$ III with C $\rightarrow$E and A $\rightarrow$C.

   (c) $\triangle$ IV $\rightarrow$ $\triangle$ III with F $\rightarrow$D, C and E interchanging.

3. (a) On a sheet of graph paper draw the lines $x = -1$ and $y = 3$.

   (b) Plot the points $(0, 2)$, $(-2, 0)$ and $(0, 0)$. Join them to make a triangle and label it T.

   (c) Reflect the triangle in the line $y = 3$ and label it T_1.

   (d) Reflect T in the line $x = -1$, and label it T_2.

   (e) What transformation maps T_2 on to T_1?

4. The points $P(-3, -2)$, $Q(1, -1)$ and $R(2, 0)$ are 3 vertices of a parallelogram PQRS.

   (a) Plot the 3 points and complete the parallelogram.

   (b) What are the co-ordinates of S?

   (c) What transformation maps SR onto PQ?

   (d) What transformation maps 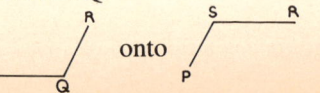 ?

5.
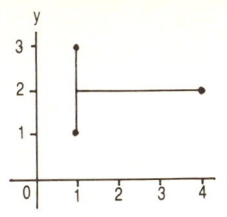

(a) Copy the shape on to a piece of graph paper, and state the equation of its line of symmetry.

(b) Rotate the shape 180°, about the origin.

(c) Another position of the original shape is given by the co-ordinates $(-1, 1)$, $(-1, 3)$ and $(-4, 2)$. State which transformation maps this position onto the original.

(d) State, without further drawing, which transformation maps the original onto $(5, 5)$, $(5, 15)$ and $(20, 10)$.

6.   The diagram below shows an object O reflected in ℓ_1 and then its image in ℓ_2.

(a) Draw accurately two parallel lines ℓ_1 and ℓ_2, 9cm apart and place the object 3cm from ℓ_1. Reflect the object first in ℓ_1 and then in ℓ_2; measure OO''.

(b) Suppose the object is first reflected in ℓ_2 and then in ℓ_1, state in words where the final image will be.

(c) Place a dot of your own somewhere *between* the two lines ℓ_1 and ℓ_2; reflect the dot first in ℓ_1 and then in ℓ_2; comment on its final location.

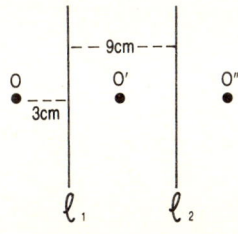

(NOT DRAWN ACCURATELY)

MIXED EXAMPLES 1
TIME 1½ HOURS

1. Solve the equations:

    (a) $\dfrac{x - 3}{3} = x$;

    (b) $\dfrac{1 - x}{3} = 7$.

2. (a) Given that $p = mq - rs$, rearrange the formula to make r the subject.

    (b) Given that $p = m^3 - r^3$, calculate p when $m = r = -1$.

3. 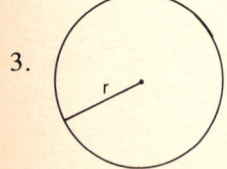 The area of this circle is 60cm²; calculate its radius.

4. 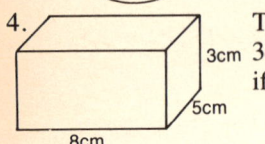 The measurements of a box are 8cm × 5cm × 3cm; by what percentage will the volume increase if each side is increased by 1cm?

5. (a) Draw accurately a square with diagonals 7cm long.

    (b) Construct a triangle of area 7½cm², marking relevant measurements. The triangle must not be right angled.

6. The scale on a map is 1:150000.

    (a) How far apart in km are two places which are 3cm apart on the map?

    (b) A reservoir on the map takes up 6cm²; what is its actual area?

7. 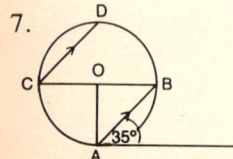 TA is a tangent to the circle centre O; CD is parallel to AB.

    Calculate   (a) DĈB   (b) DB̂C.

8. Calculate the size of each interior angle of a regular octagon.

9. A stallholder buys 30 pineapples from a fruitmarket for £19.50; she ''marks-up'' each pineapple by 40%. How much would a customer pay for 5 pineapples at this stall?

10. The intake of First Years in a school fell from 105 to 84; express this decrease as a percentage.

11. Divide £42 between 3 children so that the middle-aged child gets double what the youngest gets but only half of what the oldest gets.

12. Calculate the final bill for 746 units @ 4.4p each plus VAT at 15%.

13. (a) Which of the following are prime numbers: 19, 213, 625, 1092?

(b) By converting $^{24}/_{25}$ to a decimal or otherwise calculate $\sqrt{1^{24}/_{25}}$, giving your answer as a *fraction*.

14. Find the next number in this sequence: $1^1/_{10}$, $1^1/_2$, $1^9/_{10}$, _____.

15. Two dice are thrown; what are the chances the sum of the scores is a prime number?

16. (a) The mean of 3, 4, x, 9 is 12; calculate x.

(b) If the median of 1, 9, y, 4, 3 is y, what range of values can y take?

17. 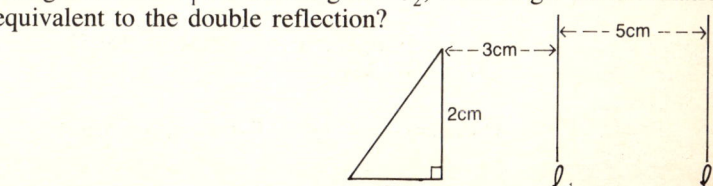 Write down Sin, Tan and Cos of angle C.

18. A security door can only be opened to a maximum of 36°; calculate how far apart the points A and B are when the door is fully opened.

19. Copy the sketch accurately with ℓ_1 parallel to ℓ_2. Reflect the triangle first in ℓ_1 and its image in ℓ_2; what single transformation is equivalent to the double reflection?

20. Copy and complete this statement (you may draw a sketch diagram to help).

If you reflect an object in the y-axis, and the image in the x-axis, the combined reflection is equivalent to a . . . .

MIXED EXAMPLES 2

TIME 1½ HOURS

1. Given that $y = \dfrac{-6}{x}$ copy and complete the table of values for (x, y).

x	−3	−2	−1	1	2	3
y			6	−6		

Plot the co-ordinates and join them to make two smooth 'half-curves'. On the graph paper draw the line $y = -x$, and estimate $\sqrt{6}$ from your graphs.

2.

x	0	1	2	3
y	6	4	2	0

The table represents a straight line; what is its equation?

3. 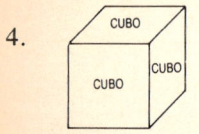 The diagram shows a rectangular area of grass surrounded by a ½m wide concrete walk; what is the area of the concrete?

4. 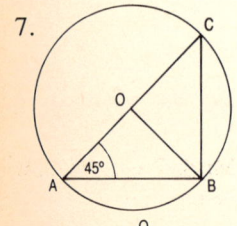 The total surface area of the cube is 384cm²; calculate the volume of the cube.

5. Two ships leave a port X, and sail on bearings of 030° and 120° respectively; if the first ship travels 120km and the second one 160km, find by an accurate scale drawing how far apart they are at the end of their routes.

6. Give the back-bearings of: (a) 033°, (b) 132°, and (c) 270°.

7. 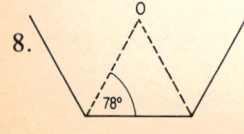 In the diagram AOC is a diameter;

(a) Calculate the size of angle C.

(b) State how many similar triangles are in the diagram.

8. The diagram shows part of a regular polygon, 'centre' O. Calculate how many sides it has.

9. (a) A special paint needs 4 tins of red mixed with every 2½ tins of green; a large drum holds 52 tins of the mixture. How many tins of each paint are needed to fill the drum?

(b) One tin of the mixture contains 300 ml; what surface area could be painted by using the whole drum, assuming 1 litre covers 13m².

10. A and B started a small company; A invested £8000 and B put in £12000; they agreed that 25% of any profits should be used for development and the remainder divided between them in the ratio of their investment.

If they made £20000 profit in the first year, how much was each partner paid out of the profits?

11. In a recipe *FOR 6 PEOPLE* two ingredients are:

30g of butter
4½lb of potatoes

Rewrite these to *SERVE 8 PEOPLE*

12. A car goes 80km an hour; convert this speed to m per second.

13. (a) Arrange the numbers ⅜, $^2/_7$, ⅓ in ascending order.

(b) Write down the next numbers in the sequence:
81, 27, 9, 1, ____, ____.

14. (a) What is the HCF and the LCM of 18 and 24?

(b) Express 210 as the product of prime factors.

15.

Mark	8	9	10
Freq.	12	10	3

The table gives the results of a French spelling test;

(a) What was the median mark?

(b) On a bar chart the column representing Mark 9 was 7½cm; how high was the column representing mark 8?

16. There are 4 girls and 5 boys eligible for Head Prefect, and Deputy Prefect; if two names are drawn at random, what are the chances they are both girls?

17.

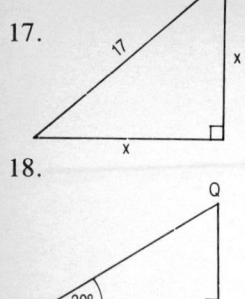

Calculate the length of x in the diagram.

18.

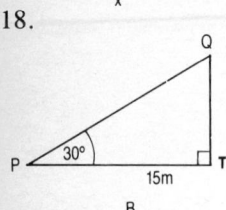

The angle of elevation of the top of a flag-post QT from a point 15m from its base is 30°;

(a) Calculate the length of the flagpost.

(b) What is the angle of depression of P from a point halfway up the post?

19.

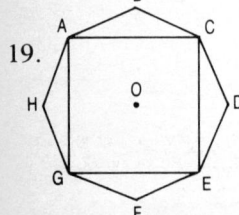

ABCDEFGH is a regular octagon; what transformation will map:

(a) △ ABC → △ GFE with A →G and C →E?

(b) △ CDE → △ AHG with C →G and E →A?

20. (a) What are the co-ordinates of the image of (2, 3) when reflected in y = −1?

(b) If (2, 3) is reflected on to (−4, 3), state the equation of the mirror line.

PART IV

ANSWERS TO PRACTICE PAPERS
CONTAINED IN PART III

ALGEBRA AND GRAPHS

1. (a) $4\frac{1}{3}$ (b) $3\frac{1}{2}$

2. (a) $x = 0$, $y = 4, 2$ (b) $-\frac{2}{3}$

3. (a) Cuttalawn: 60, 65, 70, 75, 80
   Bettalawn: 40, 50, 60, 70, 90

   (c) $20m^2$ (d) £8   (e) 32, 8

4. (a) $3x(x + 1)$ (b) $6x^2 - 18xy$ (c) $2x - 5y$

5. (a) $4\frac{1}{4}$ (b) $\dfrac{cz - ax}{b}$

6. (a) $-8, -4\frac{1}{2}, -2, -\frac{1}{2}, -4\frac{1}{2}$ (c) -4

   (d) ± 1.8 (approx.)

7. $-4x + 17y$

8. $\dfrac{qx}{100} + 5r$ or $\dfrac{qx + 500r}{100}$

AREA AND VOLUME (Answers may vary slightly)

1. (a) $3200cm^2$ (b) $1300cm^2$

2. (a) $750cm^3$ (b) $\frac{3}{4}$

3. (a) 2025   (b) 11.7   (c) $8.1m^2$ (d) £57.60

4. (a) $3\frac{3}{8}$ (b) 192   (c) $468cm^2$

5. (a) 265m   (b) over 11   (c) 34

6. (a) $615\frac{1}{2}m^2$ (b) $612cm^2$

7. $2\frac{1}{4}cm$

8. $6240cm^2$

CONSTRUCTIONS (including bearings and scales)
(Answers may vary slightly)

1. (b) 36.9°   (c) 4.8cm

2. (a) 36 miles   (b) 150°   (c) 247°

3. (b) 293°

4. (a) 165km   (b) 14mm

5. (b) 2hrs 24mins   (c) 140°

6. 1.8; $2^2/9$

7. (a) 310°   (b) 189°   (c) 120°   (d) 20°

8. (a) 1:750000   (b) 5.625   (c) 18.75

EVERYDAY ARITHMETIC 1

1. (a) £544   (b) £60, £50

2. (a) 1750, £30625   (b) 40%   (c) £6125

3. (a) 105600   (b) 88000   (c) 7%

4. (a) £140   (b) £764   (c) 9%

5. (a) 370   (b) 11   (c) 20309   (d) £1.80

6. (a) £52   (b) £24000

7. ¾pt (450ml); 24oz (681)

8. (a) 60, 120, 240   (b) £3.75, £11.25   (c) £45

EVERYDAY ARITHMETIC 2

1. (a) $2^2/5$, $8^4/5$, $17^3/5$ (b) 455

2. (a) 872   (b) LON−DOV−EXE   (c) 4hrs 36mins

3. (a) 50.4   (b) £12.50   (c) 504

4. (a) £1347   (b) 20 JUL − 26 JUL   (c) 2255

5.   (a) 80   (b) 11.36   (c) 10.27

6.   (a) £3.75   (b) 20

7.   57p; 11; 13; £1.56

8.   (a) 2hrs   (b) 10mins

GEOMETRY

1.   (a) 120°   (b) 30°   (c) 6

2.   (a) i and iii   (b) various

3.   (a) 30°   (b) 30°

4.   (a) 35°   (b) 110°

5.   (a) 66⅔   (b) 100m   (c) 896m^2

6.   (a) various   (b) various

7.   (a) 60°   (b) 80°

8.   (b) (i) Rectangle   (ii) Rhombus

NUMBERS AND NUMBER PATTERNS

1.   (a) 1, 64   (b) (i) 17   (ii) 480:16

2.   (a) 1.3x, 2x, 5x   (b) 4.987×10^{-2} (c) 1.7×10^1

3.   (a) 11.24   (b) 1p.m.

4.   (a) 6, 10, 15   (b) 10   (c) no

5.   (a) $^1/_9$, $^1/_{16}$ (b) −4, −1, 29   (c) ½, 121½
    (d) 11, 17

6.   (a) 78.25/78.75   (b) 9.3×10^7

7.   (a) 3000   (b) 30   (c) 10cm

8.   (a) 50%, 0.501, $(^5/_7)^2$ (b) 8175 or 1875
    (c) 2 + 7 + 11 (e.g.)

PROBABILITY AND STATISTICS

1. (a) $^1/_{36}$ (b) $^1/_6$

2. (a) 540   (b) 1500   (c) $20-30$ (d) $30-40$

3. (a) 80   (b) 3

4. (a) ¾ %   (b) 10(9.8)
    (c) One bar would be very long relative to the others.

5. (a) ⅓   (b) ⅓

6. (a) 5   (b) 5   (c) 4.9

7. (b) (i) 48   (ii) 108

8. (a) $^1/_{13}$ (b) $^4/_{13}$ (c) $^1/_{221}$

PYTHAGORAS AND TRIGONOMETRY

1. (a) 11.56m   (b) 85.3°

2. (a) 6.7   (b) 1.7   (c) 5.0

3. (a) 1.01m   (b) 1.09

4. (a) 60°   (b) 8.66   (c) $260cm^2$

5. (a) 18.4   (b) 90°

6. (a) 30m   (b) 36.9°   (c) 233.1°

7. (a) 26m   (b) 52.5°

8. 122m

TRANSFORMATIONS

1. (c) 1½, 6, 1:4

2. (a) Reflection in DC   (b) Rotation 90° about D
    (c) Rotation 180° about the midpoint of CE.

3. (d) Rotation 180° about $(-1, 3)$

4. (b) $(-2, -1)$

(c) Translation $\begin{pmatrix} -1 \\ -1 \end{pmatrix}$

(d) Halfturn about the midpoint of PR.

5. (a) y = 2

(c) Reflection in y-axis

(d) Enlargement scale-factor 5 about (0, 0)

6. (a) 18cm   (b) 18cm to the left of O   (c) 18cm to the right.

MIXED EXAMPLES 1

1. (a) $-1\frac{1}{2}$ (b) -20

2. (a) $\dfrac{mq - p}{s}$ (b) 2

3. 4.37

4. 80%

5. Diagrams

6. (a) 4.5   (b) $13.5km^2$

7. (a) 55°   (b) 35°

8. 135°

9. £4.55

10. 20%

11. 6, 12, 24

12. £37.75

13. (a) 19   (b) $1^2/_5$

14. $2^3/_{10}$

15. $^5/_{12}$

74

16. (a) 32   (b) $3 \leq y \leq 4$

17. 8/17, 8/15, 15/17

18. 74cm

19. Translation 10cm to right

20. Rotation 180° about (0, 0)

MIXED EXAMPLES 2

1.   (a) 2, 3, -3, -2 (b) ± 2.4

2. $2x + y = 6$

3. $20m^2$

4. $512cm^3$

5.   200km

6.   (a) 213°   (b) 312°   (c) 90°

7.   (a) 45°   (b) 3

8.   15

9.   (a) 32, 20   (b) $202.8m^2$

10. £6000, £9000

11. 40; 6

12. $22^2/_9$ m/s

13. (a) 2/7, 1/3, 3/8   (b) 1/3, 1/9

14. (a) 6, 72   (b) $2 \times 3 \times 5 \times 7$

15. (a) 9   (b) 9cm

16. $^1/_6$

17. 12.0

18. (a) 8.66   (b) 16.1°

19. (a) Reflection in HD   (b) Rotation 180° about O

20. (a) (2, -5)   (b) $x = -1$

1 litre = 1000 cubic cm

Area of triangle = $\dfrac{\text{base} \times \text{height}}{2}$

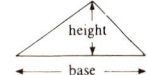

Area of parallelogram = base × height

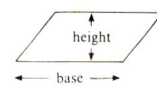

Area of trapezium = $\dfrac{1}{2}$ (a + b)h

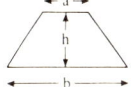

Circumference of circle = πd
$\qquad\qquad\qquad\qquad$ = 2πr

Area of circle = πr²

π = 3.14 to 2 decimal places

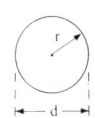

Volume of cuboid = length × width × height

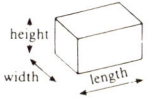

Volume of prism = area of cross-section × length

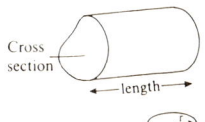

Volume of cylinder = πr²h

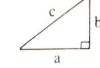

Pythagoras' Theorem
$\qquad$ a² + b² = c²

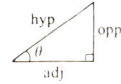

adj = hyp × cos θ
opp = hyp × sin θ
opp = adj × tan θ

or sin θ = $\dfrac{\text{opp}}{\text{hyp}}$

cos θ = $\dfrac{\text{adj}}{\text{hyp}}$

tan θ = $\dfrac{\text{opp}}{\text{adj}}$

GCSE

PRACTICE PAPERS
IN MATHEMATICS (Intermediate Level)

Teachers, parents and pupils want to know what the pupils chances are of passing at the level *entered*.

What better way of finding out than to *set aside an hour* every few weeks throughout the course of study and *test*?

There are two sets of one hour Practice Papers for each topic or group of topics (everyday arithmetic; area and volume; geometry, etc). One set can be used *during* the school term, and one for final revision.

In all, there are ten pairs of *topic based* papers *plus* two pairs of *mixed* topics (to give as close to examination requirements as possible).

NOTE:

1. The GCSE Boards Syllabi Sample and Examination Papers have been closely studied and their main requirements incorporated.

2. One hour was chosen as feasible in a *double period* at school and as a *reasonable time* to set aside for home self-testing.

DP Publications Limited
Aldine Place, 142/144 Uxbridge Road
Shepherds Bush Green
London W12 8AA
Telephone: 01 746 0044

ISBN 1-870941-30-6

9 781870 941303